SPINED ASSASSIN BUG

Most **spined** assassin bugs are just 0.5 inch (1.3 cm) long. But bigger insects stay away! The bug's spines make it hard to swallow.

Assassin bugs also have sharp beaks. They can stab prey. Their **venom** can kill other bugs.

SHARP AND SPIKY

ANIMAL WEAPONS AND DEFENSES

by Mari Bolte

CAPSTONE PRESS
a capstone imprint

Published by Capstone Press, an imprint of Capstone
1710 Roe Crest Drive, North Mankato, Minnesota 56003
capstonepub.com

Copyright © 2025 by Capstone. All rights reserved. No part of this publication may be reproduced in whole or in part, or stored in a retrieval system, or transmitted in any form or by any means, electronic, mechanical, photocopying, recording, or otherwise, without written permission of the publisher.

Library of Congress Cataloging-in-Publication Data is available on the Library of Congress website.
ISBN: 9781669078241 (hardcover)
ISBN: 9781669078197 (paperback)
ISBN: 9781669078203 (ebook PDF)

Summary: Engaging text gives readers a quick rundown of unique animals that have sharp and spiky weapons and defenses.

Editorial Credits
Editor: Mandy Robbins; Designer: Dina Her; Media Researcher: Jo Miller; Production Specialist: Tori Abraham

Image Credits
Alamy: Animal Stock, 29, Nature Picture Library, 6, Oliver Thompson-Holmes, 9, SuperStock/RGB Ventures, 26, VWPics/Kelvin Aitken, 28; Getty Images: Kevin Schafer, 21, Paul Starosta, 17, Riza Marlon, 19, Vicki Jauron, Babylon and Beyond Photography, 5; Shutterstock: Artush, 13, Black Liquid, Cover (top left), Damian Duffy, 14, Danita Delimont, 4, FJAH, 11, FoxGrafy, design element (throughout), Gena Melendrez, 25, Janelle Lugge, 15, Jesus Cobaleda, 23, JPL Designs, 18, Manuel E. Garci, 24, MarcusVDT, 7, reptiles4all, 12, Roberto Dani, Cover (bottom), Victoria Ditkovsky, Cover (top right)

Any additional websites and resources referenced in this book are not maintained, authorized, or sponsored by Capstone. All product and company names are trademarks™ or registered® trademarks of their respective holders.

Printed and bound in China. 5827

TABLE OF CONTENTS

Eat or Be Eaten 4

Spikes in the Sky 6

Thorns on the Ground 10

Sharp Sea Creatures 22

      Glossary 30

      Read More 31

      Internet Sites 31

      Index 32

      About the Author 32

Words in **bold** are in the glossary.

EAT OR BE EATEN

For many animals, it's eat or be eaten. Sharp and spiky weapons keep **predators** away. They help animals fight and find food.

Take penguins, for example. They look cute. But their tongues and the tops of their mouths are covered in spikes! The spikes keep fish from escaping as the penguin eats.

SPIKES IN THE SKY

HARPY EAGLE

Harpy eagles eat deer and snakes. They also dine on sloths and monkeys. Their **talons** can be 3 to 5 inches (7.6 to 12.7 centimeters) long. The eagles' strong legs and feet crush **prey**.

THORNS ON THE GROUND

SPANISH RIBBED NEWT

When a Spanish ribbed newt feels threatened, its ribs poke through its body. Those ribs are spiky. They can also stab an enemy with poison. Animals who try to eat the newt can get sick, or even die.

FUN FACT

The orange dots on the newt's body are where its ribs poke out.

TENREC

Tenrecs are part of the hedgehog family. Five **species** have spikes. When in danger, they curl up into a ball. No predators can get a tenrec once it's rolled itself up!

Hooked claws help tenrecs climb high.
They can go up trees and over steep rocks.
Sometimes, they hang from one foot.

THORNY DEVIL

Spikes protect the thorny devil. But it has another trick to stay safe. This lizard has a "false head" on the back of its neck. If it is being hunted, the lizard hides its real head. A predator might grab the false head instead.

FUN FACT

The false head is not an actual second head. It's a lump that fools predators.

HAIRY FROG

Male hairy frogs have skin that looks like hair. But the frog's spiky surprise is hidden in its feet. These frogs can break their toe bones. The bones poke through the skin of its toes. Hairy frogs use them to fight.

FUN FACT

The hairy frog is also called the Wolverine frog, like the X-Men character.

BABIRUSA

Babirusas are a type of pig. Lower **tusks** are used for attacking. But males also grow curved upper tusks! No one is sure why. They might be for fighting. They might be to show off. Or they might protect the animal's face.

FUN FACT

Babirusas must grind their upper tusks down. If they get too long, they can grow into the animal's skull.

GIANT ARMADILLO

The giant armadillo has the biggest claws of any animal! It uses them to dig for food. A giant armadillo can eat a whole **termite** mound in one meal.

Giant armadillos also have more teeth than any other land mammal. There are 80 to 100 chompers inside its mouth!

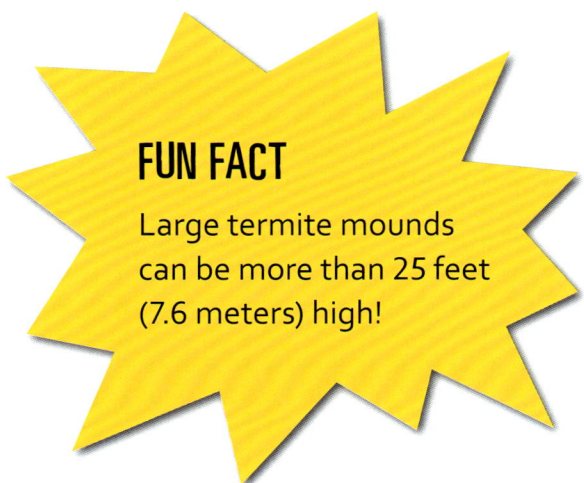

FUN FACT
Large termite mounds can be more than 25 feet (7.6 meters) high!

SHARP SEA CREATURES

SCORPIONFISH

Scorpionfish are one of the most poisonous animals in the ocean. Their bodies have between 11 and 17 slimy spikes. The slime is deadly. When the spines break another animal's skin, the slime poisons the animal.

LAMPREY

A lamprey feeds off other fish like a vampire. Its tongue scrapes away the fish's skin. Its mouth latches on like a suction cup. Teeth help it hang on too. One lamprey can kill 40 pounds (18.1 kilograms) of fish a year.

PORCUPINE RAY

The back of a porcupine ray is covered in sharp thorns. Its skin is also made up of tiny spikes. Unlike some rays, its tail does not have venom. But it is covered in extra thorns!

Porcupine rays can be as big as a kitchen tabletop. But don't worry. They only eat fish and shellfish.

GOBLIN SHARK

Goblin sharks have a mouth full of teeth. They can be seen even when the shark's mouth is closed. The shark can pop its jaw out of its mouth. It grabs prey with its teeth. Then it pops its jaw back into place.

FUN FACT
Goblin sharks have between 35 to 53 rows of teeth on their top jaw!

GLOSSARY

predator (PRED-uh-tur)—an animal that hunts other animals for food

prey (PRAY)—an animal hunted by another animal for food

species (SPEE-sheez)—a group of animals with similar features

spine (SPINE)—a hard, sharp, pointed growth such as a thorn or cactus needle

talon (TAL-uhn)—a long, sharp claw

termite (TUR-mite)—an ant-like insect that eats wood

tusk (TUHSK)—one of the pair of long, curved, pointed teeth of an animal

venom (VEN-uhm)—a poisonous liquid produced by some animals

READ MORE

Griffin, Annabel. *Weird and Wonderful Sharks*. Cornwall, UK: Hungry Tomato Ltd, 2023.

Loh-Hagan, Virginia. *Outrageous Oddities*. Ann Arbor, MI: Cherry Lake Publishing, 2023.

Norton, Elisabeth. *Deepest Divers*. Mendota Heights, MN: Apex, 2023.

INTERNET SITES

10 Fun Facts About the Harpy Eagle
audubon.org/news/10-fun-facts-about-harpy-eagle

Lamprey
kids.britannica.com/kids/article/lamprey/353359

Thorny Devil Facts for Kids
kids.kiddle.co/Thorny_devil

INDEX

babirusas, 18, 19

claws, 13, 20

giant armadillos, 20
goblin sharks, 28, 29

hairy frogs, 16, 17
harpy eagles, 6

lampreys, 24

penguins, 5
poison, 10, 22
porcupine rays, 27
predators, 4, 10, 12, 14, 15
prey, 6, 8, 28

scorpionfish, 22
skin, 16, 22, 24, 27
Spanish ribbed newts, 10, 11
spikes, 5, 12, 22, 27
spined assassin bugs, 8
spines, 8, 22

teeth, 20, 24, 28, 29
tenrecs, 12, 13
termite mounds, 20
thorny devils, 14, 15
tongues, 5, 24
tusks, 18, 19

venom, 8, 27

ABOUT THE AUTHOR

Mari Bolte is the author and editor of hundreds of children's books. Every book is her favorite book as long as the readers learned something and enjoyed themselves!